Process Intensification

Topical Conference at the 2021 AIChE Spring Meeting and 17th Global Congress on Process Safety

Online
18 – 23 April 2021

ISBN: 978-1-7138-5336-7

Printed from e-media with permission by:

Curran Associates, Inc.
57 Morehouse Lane
Red Hook, NY 12571

AIChE
120 Wall Street, FL 23
New York, NY 10005-4020

Phone: (800) 242-4363
Fax: (203) 775-5177

www.aiche.org

Additional copies of this publication are available from:

Curran Associates, Inc.
57 Morehouse Lane
Red Hook, NY 12571 USA
Phone: 845-758-0400
Fax: 845-758-2633
Email: curran@proceedings.com
Web: www.proceedings.com

TABLE OF CONTENTS

618669 An Intensified H_2 Production Process from Biogas Via Chemical Looping Water Splitting

Abstract:

H_2 is a clean and efficient secondary energy source. Although the combustion of H_2 does not emit CO_2, the production of H_2 can result in significant CO_2 emissions. Globally, more than 96% H_2 is produced from non-renewable fossil fuels including coal, oil and natural gas, which raises great concerns not only in CO_2 emissions, but also in the long-term sustainability of those processes. Biogas, a mixture primarily containing CH_4 and CO_2, is a promising alternative feedstock for H_2 production. Unlike traditional fossil fuels, biogas is both renewable and carbon-neutral. However, CO_2 removal is usually an indispensable step for H_2 production from biogas using conventional CH_4 reforming processes, lowering the energy efficiency and increasing the capital cost of the overall process.

The chemical looping water-splitting (CLWS) concept offers an alternative pathway to bypass the CO_2 removal step associated with H_2 production from biogas, resulting in significant process intensification by eliminating not only the CO_2 separation unit, but also the water gas shift unit, the H_2 purification unit, the biogas compressor and the combustion furnace. The CLWS process performs the CH_4 reforming reaction in three reactors, namely a reducer (fuel reactor), an oxidizer (steam reactor), and a combustor (air reactor). In the reducer, biogas is fully oxidized into CO_2 and H_2O by metal oxide particles. In the oxidizer, reduced metal oxide particles react with steam to produce pure H_2. In the combustor, the metal oxide particles are fully regenerated by air.

In this work, the CLWS process is simulated in ASPEN Plus to quantify its benefits compared to conventional reforming systems. The material, heat and power balances of the CLWS system are established in the process model. Sensitivity analysis on the influence of feedstock CO_2 concentration on the system performance is studied in ASPEN Plus, which shows that the CLWS system can directly process biogas feedstock with up to 50% CO_2 concentration without compromising the system performance. These significant improvements in thermal efficiency are a result of the high level of process intensification achieved by the CLWS process, which is even more remarkable when CO_2 capture is performed.

Anti-solvent precipitation in a continuous flow spinning disk reactor: A route to nanoparticles via high intensity micromixing

S. Sana, V. Zivkovic and K.Boodhoo*
School of Engineering, Merz Court, Newcastle University, Newcastle Upon Tyne, UK
*Corresponding/presenting author (kamelia.boodhoo@newcastle.ac.uk)

Nanoparticles are increasingly being developed for application in the pharmaceutical industry, more specifically in drug delivery. Starch nanoparticles have a number of industrial applications including polymer drug carrier. The present work investigates the precipitation of starch nanoparticles by contacting starch dissolved in sodium hydroxide with an ethanol antisolvent, in a spinning disc reactor (SDR). The spinning disc reactor uses surface rotation to produce extremely thin liquid film, within which improved micromixing can occur. The SDR also benefits from reduced residence times and near plug-flow characteristics. Such properties would be advantageous to solvent-antisolvent precipitation processes, were mixing is fundamental. The interactions of the disc surface topography (grooved and smooth) with conditions such as liquid flowrate, solvent-antisolvent ratio and disc speed are likely to impact on the hydrodynamics of the thin film flow. One of the objectives of this study is also to highlight and understand such interactions on the mixing and precipitation processes.

Various hydrodynamic conditions on a 30 cm diameter disc were tested in the mixing of a starch solution in 1M NaOH solvent stream with ethanol antisolvent, injected separately at different flow ratios between 1:1 and 1:9 respectively onto the centre of the rotating disc. Particle size and distribution of starch nanoparticles generated were measured via dynamic light scattering in a Malvern Nanosizer and by TEM.

The effect of disc rotational speed on the size and particle size distribution (PSD) of starch nanoparticles is displayed in Figure 1. An increase in disc rotational speed from 400 rpm to 1200 rpm leads to a decrease in mean particle size from 234 to 175 nm, with narrower PSDs. This is attributed to increased shear generating surface waves and greater instabilities within the liquid film, enhancing mixing as the liquid travels outwards on the disc surface. Similar results have been obtained for increased flow rates from 6 mL/s to 18mL/s (248 to 175 nm). Increasing the antisolvent to solvent ratio from 1:1 to 9:1 also led to a reduction in size (276 to 175 nm), as greater supersaturation was generated at reduced solubilities, causing nucleation to dominate over particle growth. The disc texture did not demonstrate a substantial impact on nanoparticle size; however, particles produced on the grooved disc were generally of narrower PSD with higher yields.

Induction times and nucleation rates have also been experimentally determined visually using a high speed video camera. Nucleation rates increased with an increase in disc rotational speed and flow rate (Figures 2 and 3) but was found to be less dependent on antisolvent to solvent ratio. A maximum nucleation rate of 6.44×10^{16} mL^{-1} s^{-1} has been estimated at conditions of 1200 rpm, 9:1 ratio and 15 mL/s, on a smooth rotating disc, whereas a maximum nucleation rate of 5.03×10^{16} mL^{-1} s^{-1} was achieved on the grooved disc for similar hydrodynamic conditions. The advantage of the grooved disc was demonstrated at conditions of low shear rates (i.e at low flow rate and disc speed) by higher nucleation rates compared to the smooth disc, highlighting the passive mixing induced via the presence of flow disrupting grooves.

Experimentally obtained nucleation kinetics along with growth kinetics have been applied to formulate a predictive mathematical model. The model combines the population balance equations (PBEs) with an engulfment micromixing model. The Hounslow's discretisation model which accounts for nucleation as well as size-dependent growth and agglomeration, has been applied to solve the PBEs and generate predictive PSDs. Figure 4 compares the modelled PSD for two different flow rates with experimental PSDs, overall showing a good agreement between the two PSDs at high flowrates. However the model does not account for film breakdown likely to occur at the low flow rate, causing the slight difference between the two PSDs at low flowrates. Figure 5 compares modelled vs experimental PSD at 400 and 1200 rpm. There is a good fit at both high and low rotational speeds.

This study, being the first comprehensive experimental and modelling investigation into solvent-antisolvent precipitation in an continuous flow SDR, has demonstrated its capability as an effective crystalliser, achieving small particles and narrow size distributions through the enhanced micromixing provided in the SDR environment, as well as providing a means to control nanoparticle properties.

Keywords. *Process intensification, spinning disc reactor, micromixing, solvent-antisolvent precipitation, nanoparticles, population balance model*

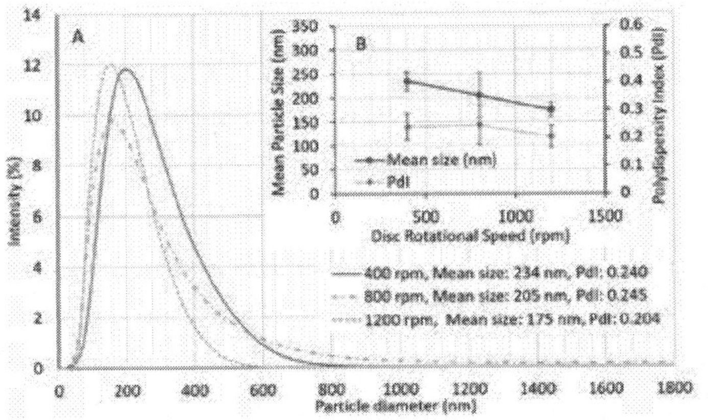

Figure 1: Effect of disc rotational speed at 18 mL/s, 9:1 antisolvent to solvent ratio, on the grooved disc A) PSD, B) plot of mean particle size and PdI.

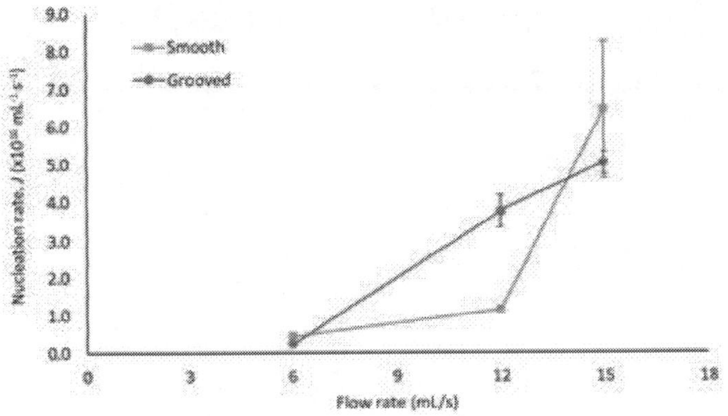

Figure 2: Effect of flow rate on nucleation rate (1200 rpm, 9:1 ratio).

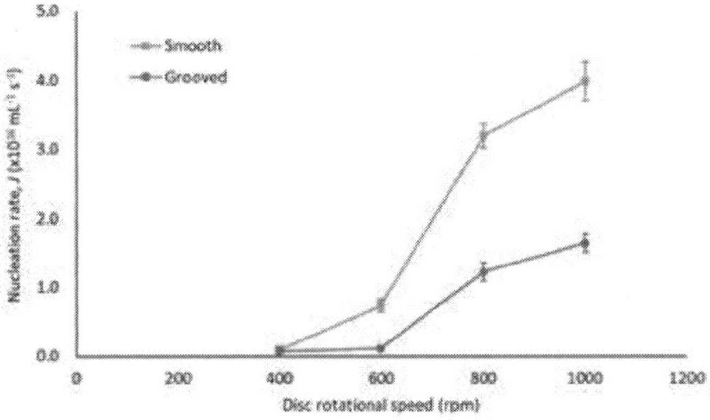

Figure 3: Effect of disc rotational speed on nucleation rate (18 mL/s, 9:1 ratio).

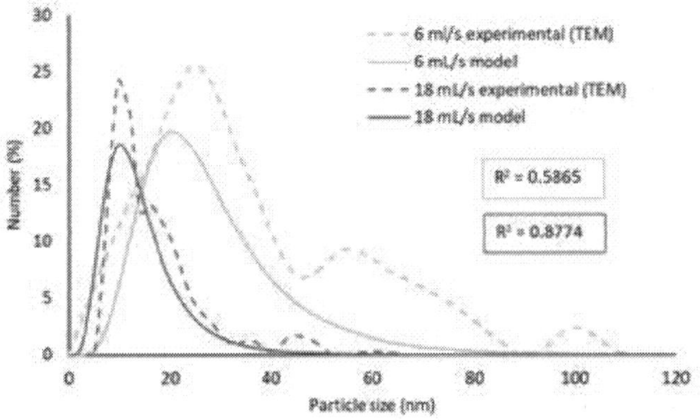

Figure 4: Effect of flow rate on experimental and modelled PSD, at 1200 rpm and 9:1 antisolvent to solvent ratio, grooved disc.

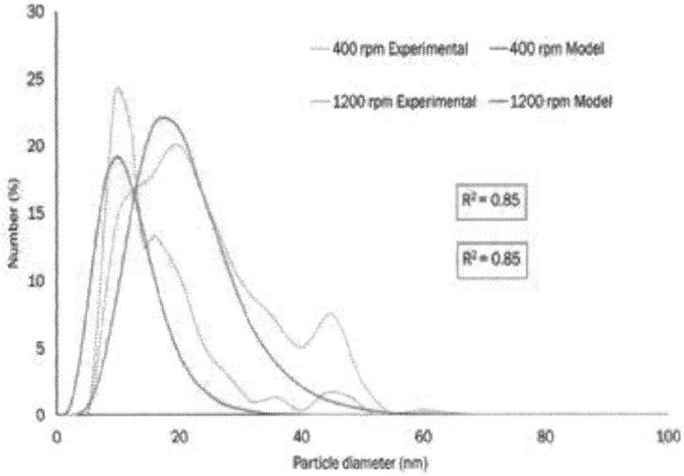

Figure 5: Effect of disc rotational speed on experimental and modelled PSD, at 18 mL/s and 9:1 antisolvent to solvent ratio, grooved disc.

Development of a microalgae biocomposite-integrated Spinning Disc Bioreactor (SDBR) for bioprocess intensification of light-driven CO_2 absorption

K.Boodhoo[1*], T. Ekins-Coward[1], S. Velasquez-Orta[1] and M. C. Flickinger[2,3]

[1] *School of Engineering, Merz Court, Newcastle University, Newcastle Upon Tyne, NE1 7RU, UK*

[2] *Chemical and Biomolecular Engineering, Engineering Building 1, North Carolina State University, Raleigh, North Carolina, 27695,USA*

[3] *Golden LEAF Biomanufacturing Training and Education Center, BTEC, North Carolina State University, Raleigh, North Carolina, 27695, USA*

*Corresponding/presenting author (kamelia.boodhoo@newcastle.ac.uk)

We report the development of an entirely new, scalable, solar energy driven microbial gas absorber-converter technology by a novel combination of advanced photoreactive biocomposite materials with a continuous thin film flow absorber design based on a spinning disc concept. A methodology was developed to integrate *Chlorella vulgaris* microalgae cells into a porous paper biocomposite for the first time with the addition of the common natural biopolymer, chitosan which acts a biopolymer bridging structure and charge neutralizer to significantly improve the integration of the *C. vulgaris* cells within the paper matrix. A comprehensive study was undertaken to assess and understand the effects of the main constituent components of "fresh" *C. vulgaris* biocomposite paper highlighted that increases in chitosan dosage, MFC content, and the total cell loading had the greatest positive impact on the photoactivity of the immobilised microalgal cells measured via a fluorescence technique. Interactions between chitosan dosage and basis weight and between cell loading and MFC content were also deemed significant. For instance, highest Fv/Fm values of the order of 0.6 were achieved with the highest total cell loading of $2.4x10^{13}/m^2$ of biocomposite surface and low MFC content (<20%) or with high chitosan levels (75 mg) and low basis weights (50 g/m^2). The microalgal cells in the biocomposite have also been shown to thrive after exposure to long periods of desiccation and shear stress where higher recovery of photosynthetic reactivity was achieved at high chitosan levels and high total cell loading.

The proof of concept of employing an immobilized *C. vulgaris* biocomposite paper on a spinning disc for CO_2 capture has been demonstrated over a period of 15 hours. Using a biocomposite paper containing 75 mg chitosan, 10% MFC, $8.8x10^{12}$ cell loading/m^2 of biocomposite surface and a basis weight of 95 g/m^2, high photoactivity was maintained at a spin speed of 300 rpm throughout the duration of operation, more so in the presence of bicarbonate in the liquid medium and 5% CO_2 in the gas environment (Experiment 1 in Table 1). The immobilised microalgae in the biocomposite preferentially consumed the dissolved CO_2 over HCO_3^-, as the dissolved CO_2 was consistently significantly lower during the SDBR runs with the *C. vulgaris* biocomposite when compared to the control blank for Experiment 1 in Figure 1. On average the liquid medium contained 54.9±0.9 % less dissolved CO_2 when compared to the blank SDBR runs. The CO_2 biofixation rate was estimated to be 144.05 mg hr^{-1} when calculated from the average difference in dissolved CO_2 per hour between the biocomposite and blank. Nayak et al. [1] reported a corresponding value of 41.6 mg L^{-1} hr^{-1} (996.4 mg L^{-1} d^{-1}) or 83.2 mg hr^{-1} for their 2 L reactor volume when simultaneously supplying supplementing 1% (v/v) CO_2 and $NaHCO_3$. Given that the result reported by Nayak et al. was from a 2 L flat-panel photobioreactor compared to the *C. vulgaris* biocomposite surface area on the disc of 62.2 cm^2, a

[1] Nayak, M., et al., Enhanced carbon utilization efficiency and FAME production of Chlorella sp. HS2 through combined supplementation of bicarbonate and carbon dioxide. Energy Conversion and Management, 2018. 156: p. 45-52.

considerable enhancement of CO_2 biofixation is evident in the present study. Interestingly, with the biocomposite paper securely fixed to the disc surface using a pH neutral spray adhesive, practically all *C. vulgaris* cells in the biocomposite successfully remained attached to the disc throughout the 15 hour rotational period at 300 rpm (equivalent to 5× g at disc edge). This is a significant achievement in the context of rotating biofilm technologies which generally must operate under gentle rotations below 20 rpm typically to prevent the microbial cells employed in wastewater treatment from detaching from the surface of biofilm structure. Overall, the increased biofixation for a much reduced spinning surface area and the high cell retention in the spinning biocomposite highlights the process intensification potential of the SDBR.

The scaling up of the biocomposite-integrated SDBR technology offers the potential of a high performance air filter, enabling many different types of microbial uptake of waste carbon gases at industrial scales such as CO_2 but also syngas (CO), biogas (CH_4), and volatile organic carbon (VOC) pollutants. With carefully selected photosynthetic micro-organisms, economically viable biotransformations of waste carbon gases into useful products such as platform chemicals and liquid fuels may also be envisaged.

Keywords: *bioprocess intensification; spinning disc bioreactor; biocomposite; microalgae; C. vulgaris; CO_2 biofixation*

Table 1. Effect of carbon source on photosynthetic activity of *C. vulgaris* biocomposite in SDBR

	Carbon source		**Maximum quantum efficiency of PSII photochemistry (F_v/F_m)**	
Experiment	**$NaHCO_3$ (gL^{-1})**	**Composition of gas**	**Initial (t=0)**	**Final (t=15 hours)**
1	2	5% CO_2 in air	0.368±0.004	0.514±0.072
2	2	Ambient Air (0.03% CO_2)	0.344±0.07	0.520±0.067
3	0	5% CO_2 in air	0.305±0.06	0.099±0.02

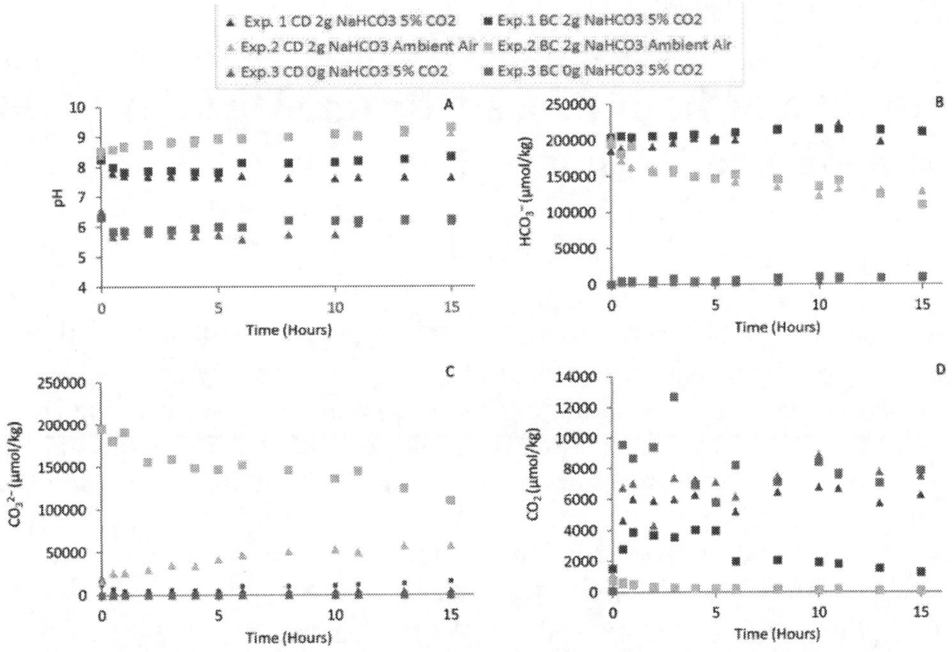

Figure 1. Results for experiments 1,2, and 3 conducted on a SDBR with a control blank disc (CD) and the *C. vulgaris* biocomposite (BC) A) The pH B) Bicarbonate (HCO_3^-) C) Carbonate (CO_3^{2-}) D) Dissolved carbon dioxide (CO_2)

619195 Paradigm Shift of Process Framework for Light Olefin Production from Shale Gas

Abstract:

The shale gas boom in the United States provides abundant natural gas liquids (NGLs) as feedstocks to petrochemical industry[1]. The major use of natural gas liquids is to produce olefins such as ethylene, propylene, and butene, which are crucial building blocks to manufacture plastic, fiber, rubber, etc. Two major bottlenecks of the conventional process framework are the conversions in reactors and the complexity of separation systems in frontend and post-reaction separation steps. Both steam cracking and catalytic dehydrogenation requires high temperatures (850°C for stream cracking and 500~650°C for catalytic dehydrogenation of propane and butane) and low partial pressure of NGL to achieve high conversions per pass. Furthermore, steam is introduced to the steam cracker to further decrease the partial pressure of hydrocarbons; diluents such as CH_4, N_2, H_2 or steam are also sometimes introduced to catalytic dehydrogenation reactor [2-5]. However, additional separation steps to remove these diluents contribute significantly to the process complexities and the costs associated with the process. Separation units in the system, such as cryogenic demethanizer, paraffin/olefin separations are conventionally achieved through complex heat-pumped distillation configurations. More importantly, several post-reaction separations are repeated separations as they are also required in the front end. Due to this duplication of separation steps, not only more energy is consumed but more equipment is employed contributing to increased capital expenditure.

Here we propose a paradigm shift of process framework, which eliminates repeated separation units and enhances conversions in the reactors synergistically. This paradigm shift of process framework distinguishes our work from all the process intensification studies on this topic in literature, which stick to the conventional process framework and focus on the development of alternative technologies and the integration of adjacent unit operations. The benefits of the paradigm shift are illustrated through detail simulations and economic analyses of exemplary processes under the new process frameworks.

Reference

1. Siirola, J. J. The impact of shale gas in the chemical industry. AIChE Journal 2014, 60, 810–819.
2. Leonard, L. E.; Bozzano, A. G.; Towler, G. P. Hydrocarbon dehydrogenation with inert diluent. 2013; US Patent App. 13/328,931.
3. Sundaram, K.; Fernandez-Baujin, J. Effect of methane and hydrogen during thermal cracking of light hydrocarbons. AIChE journal 1988, 34, 321–325.
4. Leesemann, C. J. Process for dehydrogenation of hydrocarbons in the presence of a gaseous diluent. 1949; US Patent 2,461,331.
5. Gami, A. C. Use of hydrocarbon diluents to enhance conversion in a dehydrogenation process at low steam/oil ratios. 2014; US Patent Application 14/052,949.

617477 Process Intensification of Oxidative Coupling of Methane

Abstract:

The chemical industry is faced with a dual challenge: the need for more renewable energy input and less CO_2 emissions. The combination of population growth and the increasing prosperity are consistently driving the production of chemicals, with plastics being the fastest growing group of bulk chemical products. Steam cracking of crude oil fractions, the leading production technology for the light olefins that are the key building blocks for all kinds of plastics, is very energy intensive and responsible for a large amount of CO_2 emissions, which are directly linked to the natural gas combustion in the furnace to produce the required heat input. Furthermore, on top of the natural gas that is nowadays burnt for energy and electricity generation, about 140 billion cubic meters of associated natural gas are annually flared at remote oil drilling locations around the world, accounting for 1 % of global CO_2 emissions, and this without any energy benefits. Clearly, there is an urgent need to develop economically viable processes for converting methane into more easily transportable high-value products, such as olefins and higher hydrocarbons.

Oxidative coupling of methane (OCM) is currently considered the most promising direct process for valorizing methane in the form of ethylene. OCM is initiated at the catalyst surface by the generation of methyl radicals. Methyl radicals recombine in the gas phase into ethane, which then dehydrogenates to ethylene. However, due to oxidation of the C_2 products in secondary reactions high methane conversions correspond to poor C_2 selectivities and a large yield of undesired CO_x products. Reactors with low space times and narrow residence time distribution are required to avoid oxidation of the C_2 products in secondary reactions. A second important challenge for OCM is the extreme exothermicity of the process, which makes cooling of an industrial-scale OCM reactor infeasible. To tackle these two challenges, namely the low C_2 yields due to the conversion-selectivity tradeoff and the high exothermicity, a novel intensified OCM reactor is needed. In this work, the gas-solid vortex reactor (GSVR), developed at the Laboratory for Chemical Technology (LCT) at Ghent University, is proposed for this purpose.

LCT has a cold flow, hot flow and reactive setup which provide new insights in different effects occurring at every stage of the OCM process. In combination with modeling, valuable experimental studies can be carried out for different operating conditions. Nevertheless, these time intensive experimental studies can be drastically reduced by focusing on high level computational fluid dynamics (CFD) simulations. These simulations allow to optimize the reactor geometry and operating conditions for OCM and to easily investigate alternative processes benefitting from this technology. In this project, both reactive and non-reactive CFD simulations of the GSVR are performed using the open-source CFD package OpenFOAM. The newly developed reactive CFD model takes into account a detailed OCM microkinetic model consisting of both homogeneous and heterogeneous reactions. The use of detailed kinetics is an important improvement compared to our earlier CFD work on the GSVR for OCM (presented at the EPC conference in San Antonio in 2017), where only a global 10-step mechanism was used.

CFD simulations with this newly developed framework show that the GSVR is indeed characterized by short gas-phase residence times and narrow residence time distributions. Furthermore, it is shown that the GSVR exhibits a high degree of thermal backmixing, which in adiabatic operation results in steady-state multiplicity, which in turn allows autothermal operation of the reactor. In autothermal operation the exothermic reaction heat of the OCM process is used in the best possible way, i.e., to heat the feed. Previously, autothermal operation of the GSVR for OCM was not considered, while in fact it presents the ultimate opportunity for process intensification of OCM.

618968 Carbonaceous Feedstock Conversion to Fuels or Chemicals Using High Pressure Circulating Fluidized Beds

Abstract:

SwRI has developed a technology to produce conventional liquid fuels from carbonaceous feedstocks by process intensification of industrial fuel production processes using a high pressure, high temperature, circulating fluidized bed technology. SwRI has worked with industry to develop new processes to monetize or improve the efficiency of low value petroleum and bio based feedstocks in an effort to reduce carbon dioxide emissions and improve overall economics, including projects such as the production of fuels from chicken fat and using supercritical methanol to continuously produce biodiesel. A circulating fluidized bed has the main advantages of being able to continually regenerate catalyst, maximize surface area for catalyst activity and mixing, and reduced temperature gradients. SwRI conducted small-scale experiments and is evaluating the overall fundamental technology for scale-up and process intensification as materials of construction continue to evolve to support higher pressure and temperature systems. The patented technology utilizes a fluid bed for the direct conversion of carbonaceous materials to fuels and integrates pyrolysis and deoxygenation, maximizing hydrogen usage in one single fluidized bed reactor. The high pressure reactor takes advantage of the higher partial pressures needed for hydrogenation and any benefit of supercritical conditions to work on typically low hydrogen bio derived or hydrogen depleted feedstocks. Initial testing is focused on pyrolysis and hydrotreating in a single reactor utilizing reactor design and catalyst development capabilities. This presentation will describe the high pressure circulating fluidized bed technology, future applications, and process development data.

619313 Modeling Adsorption of Organics-Laden Air from Wood Drying Using the Moving Bed Model

Abstract:

Abstract

Air leaving a wood dryer contains organics which must be removed before the air can be released to the atmosphere. The current state-of-the-art in wood drying is the regenerative thermal oxidizer (RTO), which reduces the organics to carbon dioxide over a packed bed of ceramics (and sometimes iron catalyst) at high temperatures. [1,2] Sustainability and economic pressures suggest a more passive mode of capture, where the organics are condensed and recovered instead of burned. Captis Aire has developed a moving bed thermal adsorption process which can recover organics from air using a fluidized bed adsorber. Developed as a general shortcut model for cyclical adsorption processes, the virtual moving bed model previously developed can be used to model the unit and provide insights into its performance [3,4,5] using the concept of adsorption efficiency proposed by Smith and Westerberg. [6] Pure component adsorption of representative compounds were modeled with both the Langmuir-Freundlich and thermodynamic Langmuir (tL) isotherms, [7] which were extended to mixtures using their multicomponent counterparts: the empirical loading ratio correlation (LRC) [8] and real adsorbed solution theory (RAST-aNRTL). [9,10] These isotherm predictions are used in the virtual moving bed model using the one-efficiency model, with unity bed-state efficiencies for some components, to make predictions of the performance of the fluidized bed adsorber system.

Works Cited:

1. Milota, M. R., and Western Dry Kiln Association. (2000). Regulation and Control of Air Emissions.
2. Zhang, Z., Jiang, Z., and Shangguan, W. (2016). Low-temperature Catalysis for VOCs Removal in Technology and Application: a State-of-the-Art Review. *Catalysis Today*, *264*, 270 – 278.
3. Sees, M.D., Kirkes, T., and Chen, C.-C. (2020). A Simple and Practical Process Modeling Methodology for Pressure Swing Adsorption. *submitted.*
4. Sees, M.D. (2018, October). "Novel Steady-State Process Modeling Methodology for Pressure Swing Adsorption." Paper presented at the 2018 Annual Meeting of the American Institute of Chemical Engineers, Pittsburgh, PA.
5. Sees, M.D. (2020, November). "Modeling of Gas Separations by Pressure Swing Adsorption Using a Novel Steady-State Methodology." Paper presented at the 2020 Annual Meeting of the American Institute of Chemical Engineers, San Francisco [virtual], CA.
6. Smith, O.J. and A.W. Westerberg. *Eng. Sci.*, 1991, 48(12), 2967 – 2976.
7. Chang, C.K., Tun, H., and Chen, C.-C. (2020). An Activity-Based Formulation for Langmuir Adsorption Isotherm. *Adsorption*, 2020, 26(3), 375 – 386.
8. Yang, R. T. (1997). *Gas Separation by Adsorption Processes*(Vol. 1). World Scientific.
9. Kaur, H., Tun, H., Sees, M.D., and Chen, C.-C. (2019). Local Composition Activity Coefficient Model for Mixed-Gas Adsorption Equilibria. *Adsorption*, *25*(5), 951-964.
10. Tun, H., and Chen, C.-C. (2020). Prediction of Mixed-Gas Adsorption Equilibria from Pure Component Adsorption Isotherms. *AIChE Journal*, 1 – 9.

618958 Scaleup of a Membrane to Separate Light Olefins and Paraffins

Abstract:

The separation of light olefins from paraffins is one of the most energy-intensive unit operations in petrochemicals and refining. Distillation, the current state of the art, is both capital and energy-intensive, requiring tall towers to separate ethylene from ethane or propylene from propane. Membranes have long held promise for reducing the energy and capital requirements for this separation but previously, no membrane had demonstrated sufficient stability and ruggedness for commercial scaleup and adoption. Compact Membrane Systems has developed, and is in the process of bringing to commercial fruition, a membrane for this purpose. Optiperm^{TM} is a facilitated transport membrane that combines the chemical stability of fluoropolymers with the selectivity of a silver facilitating agent. Initial applications for this process involve recovering olefin from purge streams that would otherwise be sent to fuel or recycled to a splitter. System models have been completed to demonstrate the potential of Optiperm™ to efficiently recover olefins from a variety of stream compositions, and the resulting benefits of avoiding flare or increasing splitter capacity. This talk will share the scale up progress of Optiperm^{TM} as well as details surrounding the progress on a demonstration unit for Braskem. This unit is the first installation of Optiperm™ that will simultaneously produce greater than 90% purity in both the olefin and paraffin streams. We anticipate sharing results of scale up for both spiral wound cartridges and hollow fiber cartridges. Finally, facilitated transport membranes require humidification to operate efficiently. To ensure constant humidification and avoid feed humidification on a vapor feed, CMS has developed a proprietary humidification system that utilizes liquid water directly. Results from initial testing of this novel system will also be shared.

619335 Novel Aboveground Neutralization System

Abstract:

Part of a billion-dollar project at one of the chemical manufacturing facilities in the US gulf coast, two new ion exchange (IX) trains were added with six other existing. The existing neutralization system processes regeneration wastewater from the existing IX trains. The system was experiencing problems achieving the desired pH due to inefficient design of the pit system. Existing pit system processes dilute as well as acid/caustic rich streams from IX regeneration. Due to inadequate pit volume and sensitivity of pH fluctuation with addition of acid/caustic made pit operations very inefficient. Complete regeneration of one train was not feasible. This impacted the production capacity of the demineralized water further affecting the process plant capacity. Moreover, the problem intensified after the addition of two new IX trains. Initial proposal from the plant was to add three aboveground tanks to hold larger regeneration volumes, but the major drawback was high cost and larger plot space requirement. By diverting some of dilute wastewater streams to disposal, a single above ground tank neutralization system with an eductor was proposed to save capital cost and plot space. Inadequate volume problem was solved by diverting some of the dilute streams to disposal and only neutralizing water streams with measurable acid and caustic contents. Based on rigorous literature research, there is a great need for debottlenecking methods and optimization of existing neutralization setups in the industry. Improving neutralization process, IX regeneration stream analysis, and reducing environmental impact are some of the key objective of this project.

AUTHOR INDEX